Our

by Maryellen Gregoire

Consultant:
Adria F. Klein, Ph.D.
California State University, San Bernardino

capstone
classroom
Heinemann Raintree • Red Brick Learning
division of Capstone

The sun rises every day.

The sun wakes us up.

The sun warms the fields.

It helps crops grow.

The sun warms the forest.

It helps vines grow.

The sun warms the garden.

It helps flowers grow.

The sun warms the ocean.

It helps fish grow.

It helps other animals grow.

It keeps animals warm, too.

The sun helps people grow.

It keeps people warm, too.

See you tomorrow, sun!